Did you ever see a monkey?
This animal is fun to study.
You can read about monkeys
in books. Studying monkeys is
never dull!

1

Monkeys can have short or long fur. Most of them have long tails. Many monkeys use their tails to swing. As they swing, they like to chatter loudly.

We can visit monkeys at
the zoo, but not in the wild.
Monkeys never live in the wild
in North America.

Monkeys are clever animals. These seven monkeys have been trained. They are in a circus in America.

Did you ever see a turkey?
Turkeys are fun to study too.
You can read about turkeys.
You might even see turkeys
at a farm.

Many farm turkeys live in pens. Like monkeys, turkeys make a lot of noise. Monkeys squeal and screech. What do turkeys say? Gobble, gobble!

You might see turkeys in the wild. Wild turkeys live in many parts of North America.

You can learn more about monkeys and turkeys. Find six or seven books about animals. See if they tell about monkeys or turkeys.